A K.A.P.S. Guide to Preparing for Disasters

The Family Edition

Publisher:

THGM Publishing

P.O. Box 562 Geismar, Louisiana 70734

https://www.thekapsdisasterhub.com/

Letter From the Author

Dear Family,

Over the last several decades, there has been an increase in both natural and human-made hazards. These hazards have included hurricanes, flooding, oil spills, fires, chemical spills, and pandemics. These hazards have led to an increase in the loss of life and property. As a result, families are concerned about the impact that a disaster may leave on their household.

This book is designed to provide you and your family with the tools you need to be prepared for a hazardous event. After reading and completing the activities in this book you and your family should be well-equipped to respond to a hazardous event.

This book is divided into four sections: Knowledge, Attitudes, Preparedness, and Skills. The Knowledge and Attitude section highlights various types of hazards and how to properly respond in the event that a hazard becomes a disaster. The Preparedness and Skills section provides templates and activities you can use to prepare your household.

Once you have completely worked through each section of this book I challenge you to continue reviewing the information with your family so that you all are always prepared. In the event that there is a disaster be sure to take this book with you as a guide to keep you equipped with the tools, you need to overcome the disaster.

Enjoy!
Joy Semien, Ph.D.

About the Author

Joy Semien, Ph.D. is an award-winning author, presenter, and research scientist. As a research scientist, she works to develop strategies to prepare communities for disasters.

Dr. Semien's interest in disaster research peaked at an early age living in a small rural community in Southeast, Louisiana. As a child, she witnessed chemical explosions and regularly breathed in chemicals that were released from neighboring chemical plants. She watched and listened to the stories of family members who had suffered from cancer and other health disparities as a direct result of chemical exposure.

This experience instilled in her the desire to want to further her education so she can make a difference in the world's most vulnerable communities. Joy holds a Bachelor of Science degree in Biology from Dillard University (2015) with a minor in Chemistry and a master's degree in Urban Planning and Environmental Policy from Texas Southern University (2017). She also holds a Ph.D. from Texas A&M University in Urban and Regional Sciences with a concentration on Hazard Mitigation (2024).

After pursuing an advanced education, she launched L.E.E.D. With Joy LLC to teach those working in communities how to Listen, Engage, Empower, and Drive Change (L.E.E.D.). Most recently, Dr. Joy launched The K.A.P.S. Disaster Hub as a subsidiary of L.E.E.D. With Joy LLC. Today Dr. Joy writes and develops training to prepare communities for disasters. This subsidiary seeks to encourage the growth in disaster Knowledge, Attitudes, Preparedness, and Skills (K.A.P.S.) across communities.

To learn more about Dr. Joy and L.E.E.D. With Joy, visit https://leedingwithjoy.com. To learn more about The K.A.P.S. Disaster Hub, visit the website www.thekapsdisasterhub.com, and follow @leedwithjoy & @thekapsdisasterhub on Instagram, Facebook, and YouTube.

About the Illustrator

Alexis Nunez is a Houston native and emerging visual artist currently pursuing her education at the University of Houston. Her artistic journey began in high school when she was awarded Best in Show at the Houston Rodeo Art Competition—a pivotal moment that ignited her passion for visual storytelling.

Inspired by the emotional responses her work evoked in others, Alexis discovered a deep connection between creativity and human experience. This realization has continued to fuel her dedication to the arts, leading her to explore various mediums including digital illustration and graphic design. By merging traditional techniques with modern technology, she brings imagination to life through vibrant, expressive visuals.

Since the 11th grade, Alexis has collaborated on a wide range of projects, which has played a vital role in her growth as an illustrator and graphic designer. Alexis is the proud illustrator of the entire K.A.P.S. series and continues to refine her craft and share her artistic vision with diverse audiences.

Are You Wearing your Disaster K.A.P.S?

ARE YOU WEARING YOUR DISASTER K.A.P.S?

Disasters are often one of the most sudden and unexpected challenges anyone can go through, especially families. However, If you and your family are prepared, the stress associated with the events can be somewhat alleviated.

In an effort to ensure that you are well-equipped for a hazard, I use the use acronym K.A.P.S. to represent four important terms you and your family should know: (1) Knowledge, (2) Attitude, (3) Preparedness, and (4) Skills.

1. **Knowledge:** Understanding disasters and how to respond to each one
2. **Attitude:** The belief that the way we think or perceive things can influence our decisions
3. **Preparedness:** Taking steps to adequately prepare for a disaster
4. **Skill:** Practicing what you know

KNOWLEDGE

To be prepared for a disaster you must first understand which types of hazards can impact you and your family. You can then take steps to effectively respond and potentially minimize your risk of injury and/or damage.

Important Definitions

Disaster: an incident that causes damage to human lives, infrastructure, the environment, and the economy as a result of a natural or manmade event. - National Red Cross and Red Crescent Federation

Hazard: A hazard is any source of potential damage, harm, or adverse health effects on something or someone. - OSHSA

Types of Hazards

Natural Event	Manmade Event
Flood	Industrial Explosion
Fire	Chemical Leak
Tornado	Oil Spill
Windstorm	Fire
Wildfires	Blackout
Droughts	Pandemics
Earthquakes	Terrorist Attacks
Hail	

KNOWLEDGE

The word vulnerability refers to any weakness, that you and/or your family may possess that can increase your risk of injury or damage. The more vulnerable you are the more susceptible you are to a hazardous event.

What makes you vulnerable?

Where You live, work, and play can have a big impact on how vulnerable you and your family are to hazards.

Whose Vulnerable?

Students
Elderly
Children
Caregivers
Illness/ Disability/ Immobility
Lack of Income/Savings
Lack of Disaster Training
Lack of Supplies
Lack of Friends/Families
Immigration Status
Minority Populations
Social Inequities
Proximity to the Coast
Proximity to Industrial Corridors

KNOWLEDGE

Elderly, children, students, medically ill, disabled, and/or low-income groups are most vulnerable to the impacts of hazards This is often because of things like physical immobility and the inability to access resources.

People who are low-income, lack education, and have yet to be trained in disaster preparedness can have a higher risk of vulnerability. This is because people with these characteristics may or may not have the knowledge or skills needed to respond to a hazard.

Minority groups and undocumented immigrants are also groups regularly impacted by hazards. As well as groups that live close to hazards (i.e. coastal areas & industrial corridors).

ATTITUDE

Preparing for a hazard is not just collecting supplies and making a plan. It is also about changing your attitude to ensure you and your family are mentally equipped to handle a disaster. In many cases the way a person perceives a disaster will impact their stress levels, so learning how to **S.T.O.P.** the stress before it starts is important.

Addressing Misconceptions

The first step to developing a positive attitude toward a disaster is to break all misconceptions. You can help your family do this by remembering to:

1. Never underestimate a hazardous event because it can quickly turn into a disaster.
2. Always obtain your facts from the National or Local Weather Stations.
3. Follow National/Local social media sites.
4. Approach every disaster well-prepared.
5. Always be ready to evacuate.

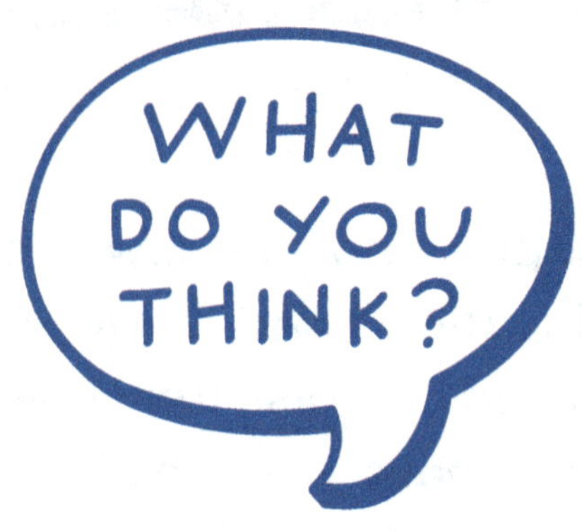

Changing Perceptions

Next, reevaluate how you perceive or think an event will occur. Remember no one event is the same. Just because you have experienced one type of event multiple times does not mean you will be okay during the next event. Stay prepared and always be ready to evacuate!

ATTITUDE

Managing Stress

Finally, learn how to control your stress level. If stress is not managed appropriately it can lead to panic and delayed responses. The more you know, the more prepared you are, and the stronger your social networks are the less stress you will endure if a hazardous event should occur.

When you are stressed remember to S.T.O.P.

Stop talking, panicking, and just breathe.
Think: about the facts, and assess the risks.
Organize: your facts and gather your belongings
Proceed: do what you believe is right.

Don't React, Just Breathe, & Respond

React: When we don't think about what to do before we make a decision this can cause us to panic.

Respond: Instead of acting without thinking take a second to gather yourself and your thoughts so you can decide how to proceed.

PREPAREDNESS

Preparing for a hazard is a "continuous cycle of planning, organizing, training, equipping, exercising, evaluating, and taking corrective action in an effort to ensure effective coordination during the incident response." - DHS, 2022.

Sheltering in Place

1. Pre-stock cabinets with water, can goods, toilet paper, etc.
2. Fill all cars with extra gas.
3. Fill all bathtubs and other big containers with water.
4. Fasten down all loose material in and around the home.
5. Board up and seal all windows and holes to the outside of the home.

Evacuating

1. Fill all cars with extra gas.
2. Bring a completed disaster supply kit for each person traveling.
3. Before leaving turn off and unplug all appliances except the refrigerator and the freezer.
4. Clean out the refrigerator and the freezer before you leave.
5. Lift up all valuable items off of the floors.
6. Update neighbors, friends, and family members on your evacuation plans.

HOW TO RESPOND TO A CHEMICAL EMERGENCY

A chemical emergency is the release of a hazardous chemical or substance. This can occur via air, water, or soil through leaks or explosions. The release has the potential to harm people's health.

How do you protect yourself and your family?

Ideally, You will <u>Shelter in Place</u> using the following steps:

- Go inside the closest building.
- Close/ lock all of the doors and windows (use tape to seal windows and other holes).
- Place a towel under the windows and doors.
- Turn off all fans, air conditioners, and heaters.
- Turn on the television/radio and tune into your local station to obtain updated alerts.
- Be sure to register with the available local emergency management list-serv and follow their social media accounts.
- Listen to a local Emergency Alert System (EAS) station for emergency instructions from the county or state officials.
- Stay inside until you hear the "All Clear" siren or are otherwise notified via phone, radio, t.v. etc.

HOW TO RESPOND TO A CHEMICAL EMERGENCY

Evacuation steps:

- Grab your disaster preparedness kit.
- Grab your disaster plan.
- Cover your nose and mouth with a towel or mask.
- Get to the car as quickly as possible.
- Ensure all windows are up, vents are closed, and the air conditioner is off!
- Drive to safety!

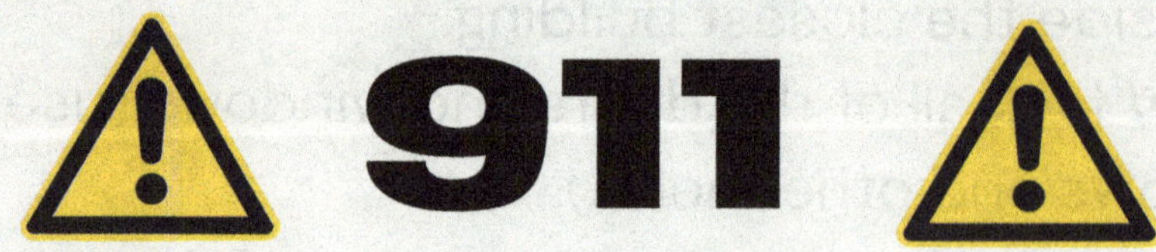

Need to Report a Chemical Emergency:

- 911
- National Response Center at 800-424-8802
- Poison Control Center (1-800-222-1222)

HOW TO RESPOND TO A HURRICANE

A Hurricane is a tropical cyclone, which forms over tropical or subtropical waters. Hurricane season is typically between June 1 and November 30.

Things to Know:

There are 5 categories of a hurricane:

- Category 5: Catastrophic Damage Wind: > 155MPH
- Category 4: Extreme Damage Wind 131-155 MPH
- Category 3: Extensive Damage Wind: 111-130 MPH
- Category 2: Moderate Damage Wind: 96-110 MPH
- Category 1: Minimal Damage Wind: 74 to 95 MPH

How do you protect yourself and your family?

Shelter in Place

- Pre-stock cabinets with water, can goods, toilet paper, etc.
- Fill all cars with extra gas.
- Fill all bathtubs and other big containers with water.
- Fasten down all loose material in and around the home.
- Board up and seal all windows and holes to the outside of the home.

HOW TO RESPOND TO A HURRICANE

How do you protect yourself and your family?

Evacuate

- Fill all cars with extra gas
- Bring a completed disaster supply kit for each person traveling
- Before leaving turn off and unplug all appliances except the refrigerator and the freezer
- Clean out the refrigerator and the freezer before you leave.
- Lift up all valuable items off of the floors
- Update neighbors, and family members on your evacuation plan

Important Reminders

- It is important to stay up to date with developing storms. Families can visit websites like www.nhc.noaa.gov, www.NOAA.gov, and adar.weather.gov to stay up to date.

HOW TO RESPOND TO A TORNADO

A tornado is "a narrow, violently rotating column of air that extends from the base of a thunderstorm to the ground." - FEMA, 2022. Remember tornado season is typically between March- July.

Things to Know:

Tornado Watch: Issued by the storm prediction center and represents a larger area. The notice is saying a tornado is possible.
Tornado Warning: Issued by the local weather center, it means a tornado has been spotted therefore take cover!

How do you protect yourself and your family?

Shelter in Place

- If in a car GET OUT and take shelter in a building or a low-lying area.
- If in a mobile home GET OUT! Take shelter in a building with a sturdy foundation.
- If in a home with a cement, foundation go into a room with no windows (hallway or closet) or the lowest area of the home.
- Place a mattress over you or another durable piece of furniture to resist debris.
- "Crouch down on your knees and protect your head with your arms."

HOW TO RESPOND TO FLOODING

Flooding is an overflow of water onto dry land. Usually the result of water filling creeks and/or river beds at a fast pace without warning. Flooding typically leads to flash floods which are extremely harmful, fast, and "unpredictable". - NOAA, 2022

Things to Know:

Flood Watch: Issued by the national weather service and represents a larger area. The notice is saying that flooding is possible. Expected to last 12 to 24 hours.

Flood Warning: Issued by the local weather center, it means flooding is occurring, therefore take shelter! Issued 24-60 hours before crest.

How do you protect yourself and your family?

- Always check the T.V. or Radio for flooding information
- Decide whether to Stay at Home or to Evacuate (If not mandatory)
- Make sure mobile or loose items are bolted to the ground & elevate all valuables
- Collect sandbags to place around homes
- Leave during daylight and check Evacuation Routes

HOW TO RESPOND TO WINTER WEATHER

Freezing indicates that ice up to 1/4 inch is expected in the area. The event may last a few hours or multiple days. - NWS, 2022

Be aware: Utilities and communication services are at risk of going out. Those at greatest risk are the elderly, children, sick individuals, and pets.

Things to Know:

Winter Storm Watch: Issued 12 to 48 hours before a winter storm. Possible blizzard, heavy snow, heavy freezing rain, or heavy sleet.

Winter Storm Warning: Issued 12 to 24 hours before a winter storm. Blizzard, heavy snow, heavy freezing rain, or heavy sleet is occurring.

Winter Weather Advisory: Issued to notify the public of unsafe weather conditions (i.e. snow, freezing rain, freezing drizzle, and sleet). Exercise extreme caution, otherwise hazard can lead to life-threatening conditions.

HOW TO RESPOND TO WINTER WEATHER

How do you protect yourself and your family?

Before the Storm

- Weatherize your home (insulate, caulk, add weather strips, and cover pipes
- Install a carbon monoxide detector
- Install a smoke detector
- Check all heating devices for electrical shortages

During the Storm

Shelter in Place

- Avoid traveling
- Stay Indoors
- Drink warm Fluid
- Heat your home safely

Evacuation

- Dress warmly: cover all body parts fully to avoid frostbite
- Take your Disaster Kit & Plan

After the Storm

- Check news outlets for emergency information
- Check all pipes
- Check all heating equipment

HOW TO RESPOND TO SUMMER WEATHER

Extreme Heat indicates that the temperature will exceed 90 degrees and the humidity will become increasingly high for multiple days.

Be Aware: Utilities and communication services are at risk of going out. Those at greatest risk are the students, elderly, children, sick individuals, those overweight, and pets.

Things to Know:

Excessive Heat Watch: Issued by the local county. This indicates a heat index is possible.

Excessive Heat Warning: Issued by the local county, 1 to 2 days before a heat wave. Indicates a heat index of over 105 °F for at least two hours.

Excessive Heat Advisory: Issued by the local county, 1 to 2 days before a heat wave. Indicates a heat index of over 95°F for at least two hours.

HOW TO RESPOND TO SUMMER WEATHER

How do you protect yourself and your family?

Before the Heat Wave

- Weatherize your home (insulate, caulk, add weather strips, and install window reflectors)
- Locate your local cooling center
- Check all cooling devices for full operational capabilities and electrical shortages
- Install insulated window air conditioning Unit

During the Heat Wave

- Take cool showers or baths
- Stay inside cool doors
- Wear loose, lightweight, and light-colored clothing
- Cool your home safely
- Drink plenty of cool liquids
- Watch for heat-related illnesses
- Check on family/friends

HOW TO RESPOND TO A EARTHQUAKE

An Earthquake is when the ground suddenly begins to shake. This is typically caused by rocks under the earth's surface shifting. - USGS, 2022

The aftermath of an earthquake can lead to fires, tsunamis, landslides, and even avalanches.

How do you protect yourself and your family?

- Drop down to your hands and Knees
- Hold on to something sturdy
- Cover your neck and your head to prevent debris from falling on you
- If possible, get under a table or something sturdy
- If outside stay away from the building and brace yourself

SKILLS

Practice makes perfect! It is extremely important for you and your family to build a disaster preparedness kit and develop a household plan! Once you have built the plan, then it is important to continuously practice the plan so you and your family are prepared for a disaster.

Steps to Build Your Skills

Design a Plan

Build a Disaster Preparedness Kit

Practice

Skills Section Instructions

The remainder of this book is designed to help you and your family build a disaster plan and a kit. As you go through the plan don't feel intimidated, just breathe and take each section one step at a time.

Suggestions for Completion:

- Turn each section of the plan into a competition. Challenge your family members to complete a section. The person who completes their section first wins.
- Challenge your children to gather supplies from around the house the child with the most items wins!
- Create a safe space that allows everyone to have an open and honest conversation. Challenge each person to share their thoughts and concerns.

Building

A Disaster Preparedness Kit

BUILDING A DISASTER PREPAREDNESS KIT

- Backpack
- Family Plan
- Small Notebook
- Pens/ Markers
- Emergency Contact Card
- Battery Packs/ Batteries
- Chargers
- Candles/ Matches
- Flashlight
- Earphones

BUILDING A DISASTER PREPAREDNESS KIT

- Non-perishable Food
- Disposable Utensils
- Disposable Bowls/Plates
- Manual can opener
- Water Purification Tablets
- 1 Gallon of Water - Per Person
- First Aid Kit (Band-Aids, antibacterial ointments)
- Radio (Battery Operated)
- Tool kit (Pocket Knife, Screwdriver, Scissors etc.)
- Cleaning Supplies

BUILDING A DISASTER PREPAREDNESS KIT

- Extra Clothes & Shoes (3 days worth per person)
- Toiletries (Toothbrush, Soap, Deodorant)
- Hand sanitizer
- Sleeping bags, Blankets, & Pillows
- Medications
- Medical Equipment
- Towels
- Toilet paper / Paper Towels
- Ziplock Bags
- Garbage Bags

BUILDING A DISASTER PREPAREDNESS KIT

- Cell Phones
- Whistle
- Duct Tape
- Sunscreen/Bug Spray
- Umbrella/ Raincoat
- Particle Respirator/N95 Mask
- Disposable Camera
- Cash & Quarters
- Gloves
- Pet Supplies

BUILDING A DISASTER PREPAREDNESS KIT

- [ ] Toys For Kids
- [ ] Diapers
- [ ] Wipes
- [ ] Formula
- [ ] ______________________________
- [ ] ______________________________
- [ ] ______________________________
- [ ] ______________________________
- [ ] ______________________________
- [ ] ______________________________

BUILDING A DISASTER PREPAREDNESS KIT

Notes

Let's Make a Plan!

Identification Cards

Using the space below tape/glue in pictures of your identification cards/ passport/green card/visa etc..

Identification Cards

Using the space below tape/glue in pictures of your identification cards/ passport/green card/visa etc..

Notes

Emergency Contact Information

EMERGENCY HOTLINE

Mobile:____________________

Telephone:____________________

Email:____________________

FIRE DEPARTMENT

Mobile:____________________

Telephone:____________________

Email:____________________

POISON CONTROL CENTER

Mobile:____________________

Telephone:____________________

Email:____________________

POLICE DEPARTMENT

Mobile:____________________

Telephone:____________________

Email:____________________

HOSPITAL EMERGENCY

Mobile:____________________

Telephone:____________________

Email:____________________

PHARMACY

Mobile:____________________

Telephone:____________________

Email:____________________

EMERGENCY CONTACT FAMILY

Mobile:____________________

Telephone:____________________

Email:____________________

EMERGENCY CONTACT FRIEND

Mobile:____________________

Telephone:____________________

Email:____________________

Important Family Members

Using the space below record the contact information for all of your important family members who can serve as an emergency contact.

Name	Phone	Address	Relationship

Important Friends

Using the space below record the contact information for all of your important friends who can serve as an emergency contact.

Name	Phone	Address

Important Organizations

Using the space below record all important organizations that you will need during an emergency.

Name	Phone	Address

Important Resources

Using the space below record the resources that you will need during an emergency.

Name	Phone	Website	Resource Type

Notes

Be Prepared

Evacuation Information

EVACUATION CHECKLIST

- [ ] Emergency Contact Card
- [ ] Medications
- [ ] Social Security Card
- [ ] Passport/ID/Visa/Green Card
- [ ] Copy of Birth Certificates
- [ ] Medical Information
- [ ] Pet Records
- [ ] Monthly Bills
- [ ] Bank Account Information
- [ ] Insurance Information (Health, Car, House)

EVACUATION CHECKLIST

Fully Packed Disaster Preparedness Kit

THINGS TO REMEMBER

When Evacuating

1. Always check the T.V. or Radio for flooding information.
2. Decide whether to Shelter in Place or Evacuate (If not mandatory).
3. Make sure mobile or loose items are bolted to the ground.
4. Elevate all valuables
5. Collect sandbags to place around homes.
6. Always check the T.V. or Radio for flooding information.
7. Leave during daylight and check evacuation routes.

When Returning Home

1. Wear protective clothing ex. gloves, long sleeves, closed-toe shoes, and pants.
2. Check for Damage to the home before entering.
3. Do not use an open flame for light.
4. Check food for spoilage.
5. Don't Drink the Tap Water.

Be Prepared

Medical Information

Doctor's Information

Using the space below record all relevant patient information.

Patient Name	Doctor's Name	Doctor's Phone #

Doctor's Information

Using the space below record all relevant patient information.

Patient Name	Doctor's Name	Doctor's Phone #

Prescription Information

Using the space below record all relevant prescription information.

Patient Name	Medication Name	Doctor (Name & Phone #	Pharmacy (Name & Phone #)

Prescription Information

Using the space below record all relevant prescription information.

Patient Name	Medication Name	Doctor (Name & Phone #	Pharmacy (Name & Phone #)

Medical Equipment

Using the space below record all relevant medical equipment information.

Patient Name	Equipment Name	Serial #	Company Contact (Name & Phone)

Medical Equipment

Using the space below record all relevant medical equipment information.

Patient Name	Equipment Name	Serial #	Company Contact (Name & Phone)

Medical Insurance Cards

Using the space below tape/glue in pictures of your insurance cards.

Medical Insurance Cards

Using the space below tape/glue in pictures of your insurance cards.

Perscription Cards

Using the space below tape/glue in pictures of your prescription cards.

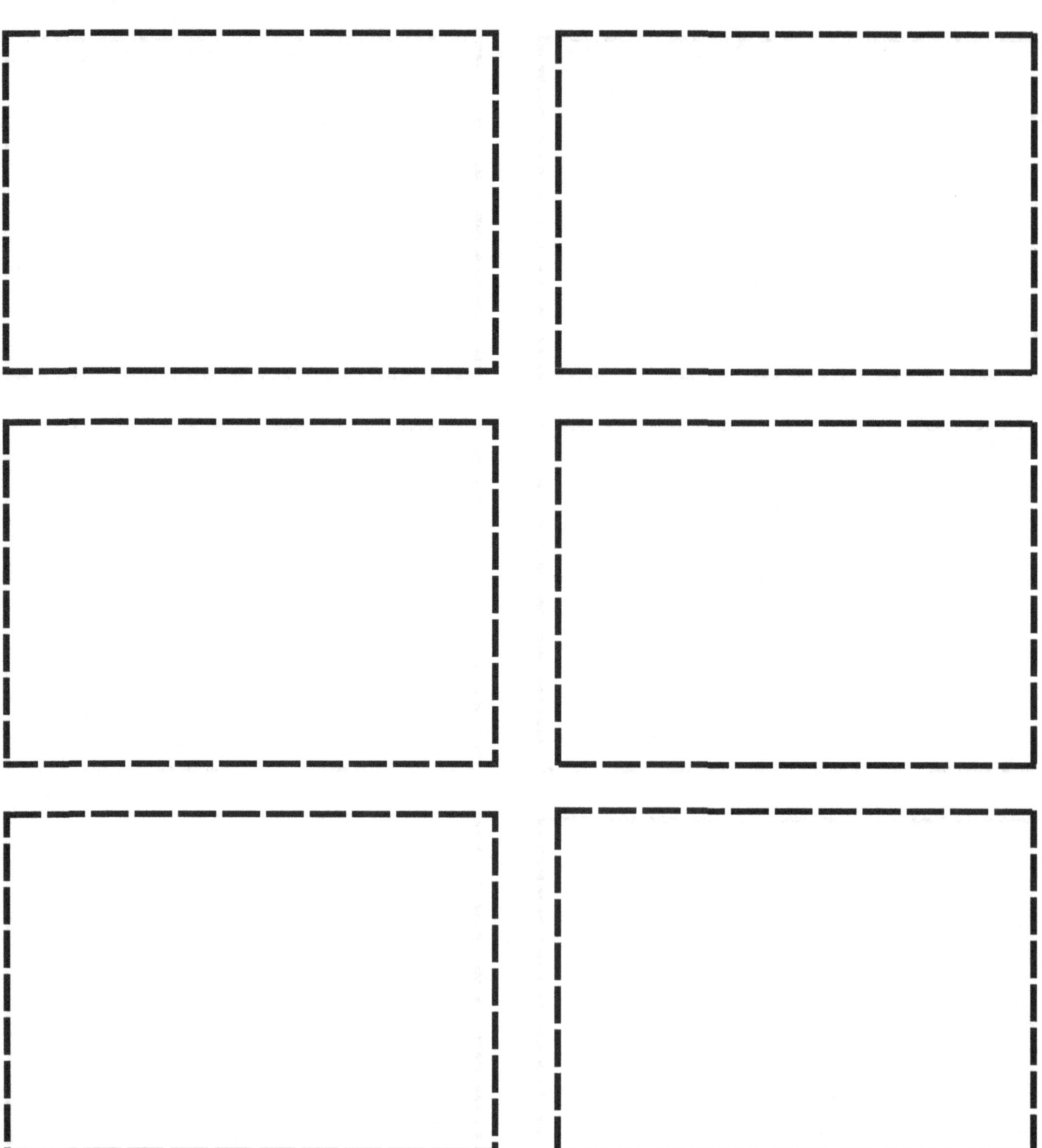

Perscription Cards

Using the space below tape/glue in pictures of your prescription cards.

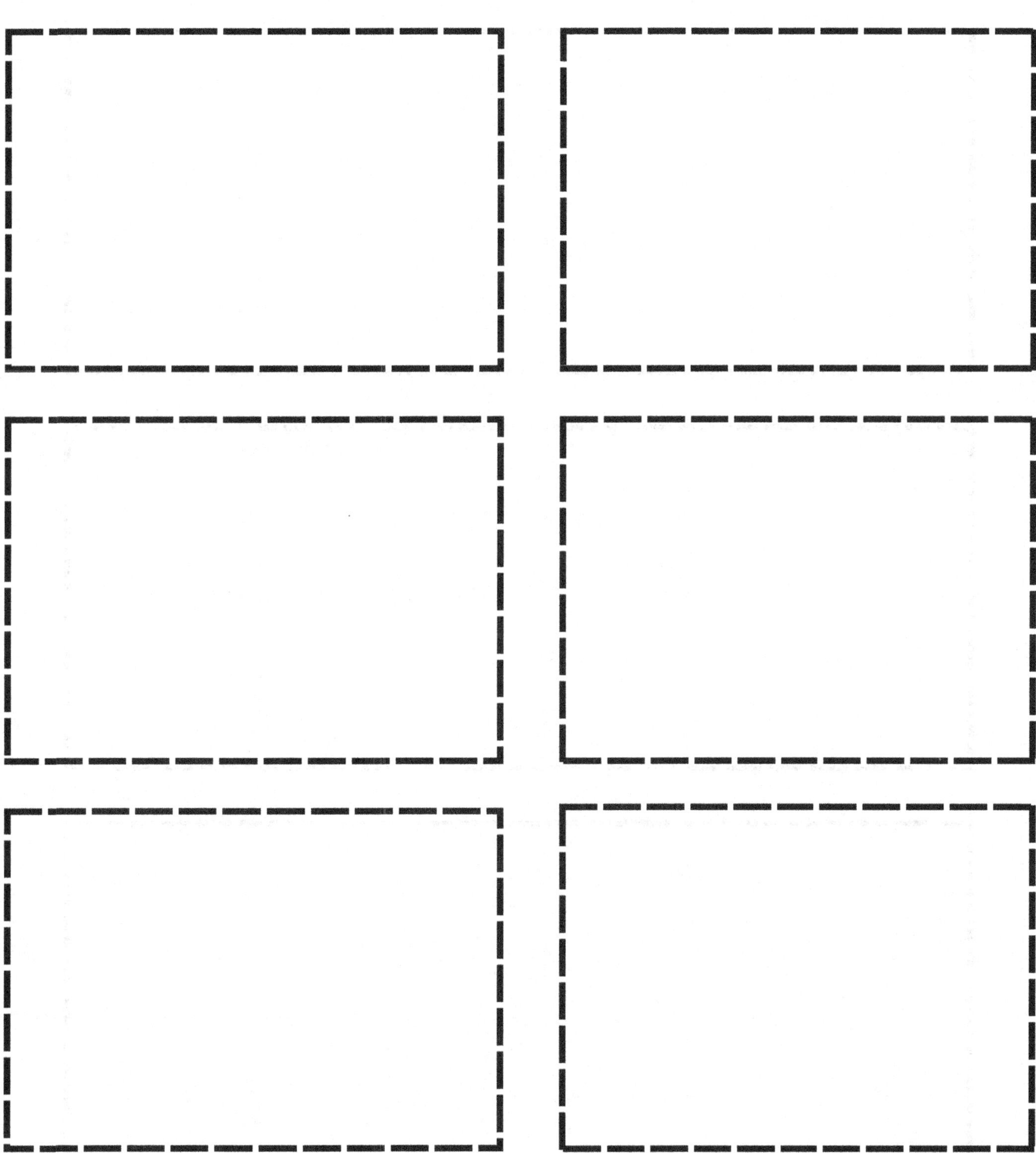

Immunization Records

Using the space below tape/glue in pictures of your immunization records

Immunization Records

Using the space below tape/glue in pictures of your immunization records

Notes

Be Prepared

Insurance Information

Insurance Information

Using the space below record all relevant insurance information.

Insurance Type	Company Name	Policy #	Agent Contact Information

Insurance Information

Using the space below record all relevant insurance information.

Insurance Type	Company Name	Policy #	Agent Contact Information

Insurance Cards

Using the space below tape/glue in pictures of your insurance cards.

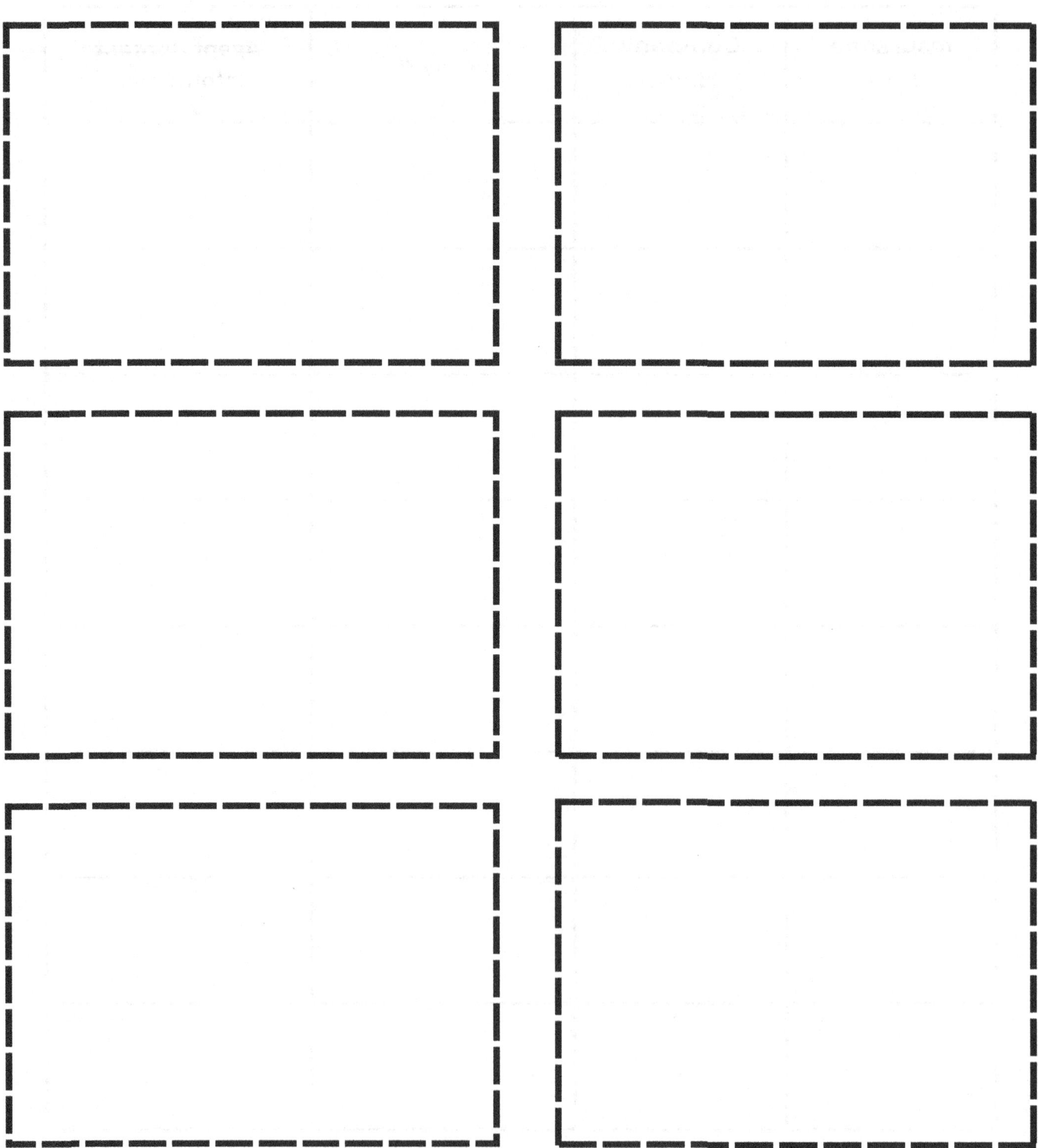

Insurance Cards

Using the space below tape/glue in pictures of your insurance cards.

Notes

Be Prepared

Finance Information

Bank Information

Using the space below record all relevant bank information.

Bank Name	Phone # /Address	Account Type	Website	Username/ Password (Hint only)

Bank Information

Using the space below record all relevant bank information.

Bank Name	Phone # /Address	Account Type	Website	Username/ Password (Hint only)

Monthly Bills

Using the space below record all relevant monthly bills.

Bill	Company	Account #	Phone #	Due Date

Monthly Bills

Using the space below record all relevant monthly bills.

Bill	Company	Account #	Phone #	Due Date

Notes

Be Prepared

Vehicle Information

Vehicle Information

Using the space below record all relevant vehicle information.

Vehicle	VIN	Owner Name	Insurance Agency	Policy #

Vehicle Title

Using the space below tape/glue in pictures of your vehicle title.

Vehicle Title

Using the space below tape/glue in pictures of your vehicle title.

Be Prepared

Education Information

School Information

Using the space below record all relevant school information.

Student	School	Address	Emergency Hotline

Transcripts

Using the space below tape/glue in pictures of your transcripts.

Transcripts

Using the space below tape/glue in pictures of your transcripts.

Transcripts

Using the space below tape/glue in pictures of your transcripts.

Transcripts

Using the space below tape/glue in pictures of your transcripts.

Transcripts

Using the space below tape/glue in pictures of your transcripts.

Transcripts

Using the space below tape/glue in pictures of your transcripts.

Be Prepared

Pet Information

PET CHECKLIST

- [ ] Adoption Records and Registration for Each Pet
- [ ] Vaccination Records
- [ ] Medication Tracker
- [ ] Picture of You and Your Pet
- [ ] Food & Water
- [ ] Bowls
- [ ] Leashes
- [ ] Pet Carriers
- [ ] ______________________________
- [ ] ______________________________

PET CHECKLIST

Pets

Using the space below record all relevant pet information.

Pet Name	Vet Name	Vet Contact (Phone #)	Pet Shelter/ Boarding Facility

Pet Records

Using the space below tape/glue in pictures of your pet records.

Notes

Be Prepared

Home Inventory

Outside Inventory

Make an inventory of all items located outside of your home.

Item	Serial Number	Manufacturer	Warranty Info.

Outside Inventory

Make an inventory of all items located outside of your home.

Item	Serial Number	Manufacturer	Warranty Info.

Outside Pictures

Using the space below tape/glue in pictures of items outside your home.

Outside Pictures

Using the space below tape/glue in pictures of items outside your home.

Electronic Inventory

Using the space below make an inventory of all electronics in your home.

Item	Serial Number	Manufacturer	Warranty Info.

Electronic Inventory

Using the space below make an inventory of all electronics in your home.

Item	Serial Number	Manufacturer	Warranty Info.

Electronic Pictures

Using the space below tape/glue in pictures of your electronics.

Electronic Pictures

Using the space below tape/glue in pictures of your electronics.

Appliance Inventory

Using the space below make an inventory of all appliances' in your home.

Item	Serial Number	Manufacturer	Warranty Info.

Appliance Inventory

Using the space below make an inventory of all appliances' in your home.

Item	Serial Number	Manufacturer	Warranty Info.

Appliance Pictures

Using the space below tape/glue in pictures of your appliances.

Appliance Pictures

Using the space below tape/glue in pictures of your appliances.

Furniture Inventory

Using the space below make an inventory of all furniture in your home

Item	Serial Number	Manufacturer	Warranty Info.

Furniture Inventory

Using the space below make an inventory of all furniture in your home

Item	Serial Number	Manufacturer	Warranty Info.

Furniture Pictures

Using the space below tape/glue in pictures of your furniture.

Furniture Pictures

Using the space below tape/glue in pictures of your furniture.

What's in Your Fridge?

Using the space below make an inventory of the food in your fridge.

Item	Expiration Date

What's in Your Fridge?

Using the space below make an inventory of the food in your fridge.

Item	Expiration Date

Fridge Pictures

Using the space below tape/glue in pictures of the items you will be leaving in your fridge upon evacuation.

Helpful Hint:

Before leaving your house. Freeze a cup of water. Place a coin on top of the frozen water. When you return if the coin is below the level at which you left it you now know the power went out and you should dispose of all your food.

Fridge Pictures

Using the space below tape/glue in pictures of the items you will be leaving in your fridge upon evacuation.

What's in Your Freezer?

Using the space below make an inventory of the food in your freezer.

Item	Expiration Date

What's in Your Freezer?

Using the space below make an inventory of the food in your freezer.

Item	Expiration Date

Freezer Pictures

Using the space below tape/glue in pictures of the items you will be leaving in your freezer upon evacuation.

Helpful Hint:

Before leaving your house. Freeze a cup of water. Place a coin on top of the frozen water. When you return if the coin is below the level at which you left it you now know the power went out and you should dispose of all your food.

Freezer Pictures

Using the space below tape/glue in pictures of the items you will be leaving in your freezer upon evacuation.

Notes

Notes

Notes

Notes

Notes

Let's Practice

Scenario

Are you prepared?

Practice Scenario 1

It's 10:30 P.M. you and your family of four have been in bed for at least an hour listening to the night rain. All of a sudden you hear someone knocking on the door. As you rise out of bed to see who is knocking on your door, you step down into a puddle of water. A little annoyed but not overly concerned you continue to the door to see who is disturbing your family's rest. As you open the door your neighbor begins to scream "The neighborhood is flooding! We have to evacuate now!" What would you do next?

1. What **type of disaster** is impacting you and your family?

   ____

2. Is this a **human-made** or **natural disaster**?

   ____

3. How would you **React** to this scenario? Why?

   ____

   ____

4. How would you **Respond** to this scenario? Why?

   ____

   ____

5. Do you **Evacuate** or do you **Shelter in Place**? Why?

   ____

6. What **items will you take with you**? Why?

   ____

   ____

   ____

   ____

7. Name ways to **stay calm** during this scenario?

   ____

   ____

   ____

   ____

Are you prepared?

Practice Scenario 2

It's 2:00 P.M. you just picked your kiddos up from school. You decide that you want to pick up some ice cream for yourself and the kids. After picking up the ice cream you turn the corner and realize the clouds are getting gray. So you decide to go straight home. By the time you get home, the sky is so dark and the rain is so thick you can barely see. All of a sudden your cell phone begins to go off with weather alerts! The alert says "Take Cover Tornado Warning!" <u>What would you do next?</u>

1. What **type of disaster** is impacting you and your family?

   ____

2. Is this a **human-made** or **natural disaster**?

   ____

3. How would you **React** to this scenario? Why?

   ____

   ____

4. How would you **Respond** to this scenario? Why?

   ____

   ____

5. Do you **Evacuate** or do you **Shelter in Place**? Why?

   ____

6. What **items will you take with you**? Why?

   ____

   ____

   ____

   ____

7. Name ways to **stay calm** during this scenario?

   ____

   ____

   ____

   ____

Are you prepared?

Practice Scenario 3

It's 5:30 A.M. you were just about to wake up and get the kids dressed for school. All of a sudden you smell smoke. You get up and look around but you don't see anything. You hop back in bed next to your partner, but the smell gets stronger so you nudge your partner to wake up. In less than two minutes the whole house is covered in smoke. What would you do next?

1. What **type of disaster** is impacting you and your family?

   ____

2. Is this a **human-made** or **natural disaster**?

   ____

3. How would you **React** to this scenario? Why?

   ____

   ____

4. How would you **Respond** to this scenario? Why?

   ____

   ____

5. Do you **Evacuate** or do you **Shelter in Place**? Why?

   ____

6. What **items will you take with you**? Why?

   ____

   ____

   ____

   ____

7. Name ways to **stay calm** during this scenario?

   ____

   ____

   ____

   ____

Draw Your Evacuation Plan

Are you prepared?

Practice Scenario 4

It's Sunday Morning, and you and your family just got home from a long Disney vacation. You decided to call your sister while you unpack your suitcase. You turn the T.V. on and all of a sudden you hear the weatherman announce that a hurricane is headed toward you and your family. The Hurricane will arrive in two days and it's expected to be a Category 4. <u>What would you do next?</u>

1. What **type of disaster** is impacting you and your family?

   ____

2. Is this a **human-made** or **natural disaster**?

   ____

3. How would you **React** to this scenario? Why?

   ____

   ____

4. How would you **Respond** to this scenario? Why?

   ____

   ____

5. Do you **Evacuate** or do you **Shelter in Place**? Why?

   ____

6. What **items will you take with you**? Why?

   ____

   ____

   ____

   ____

7. Name ways to **stay calm** during this scenario?

   ____

   ____

   ____

   ____

Are you prepared?

Practice Scenario 5

It's Saturday morning, you and your family are sitting at the table eating pancakes for breakfast! You're enjoying your meal. All of a sudden you hear the chemical siren going off alerting you that there is a chemical emergency in your neighborhood. <u>What would you do next?</u>

1. What **type of disaster** is impacting you and your family?

   ____

2. Is this a **human-made** or **natural disaster**?

   ____

3. How would you **React** to this scenario? Why?

   ____

   ____

4. How would you **Respond** to this scenario? Why?

   ____

   ____

5. Do you **Evacuate** or do you **Shelter in Place**? Why?

   ____

6. What **items will you take with you**? Why?

   ____

   ____

   ____

   ____

7. Name ways to **stay calm** during this scenario?

   ____

   ____

   ____

   ____

Are you prepared?
Practice Scenario 6

It's Tuesday Afternoon, you and your family are watching the afternoon news. You hear the weatherman announce "Expect a winter freeze for the rest of the week. The roads will be covered in ice, snow will be falling, and there is a strong possibility that the power will be out." What would you do next?

1. What **type of disaster** is impacting you and your family?

 ____

2. Is this a **human-made** or **natural disaster**?

 ____

3. How would you **React** to this scenario? Why?

 ____

 ____

4. How would you **Respond** to this scenario? Why?

 ____

 ____

5. Do you **Evacuate** or do you **Shelter in Place**? Why?

 ____

6. What **items will you take with you**? Why?

 ____

 ____

 ____

 ____

7. Name ways to **stay calm** during this scenario?

 ____

 ____

 ____

 ____

Are you prepared?

Practice Scenario 7

It's Wednesday Morning, you and your family are watching the morning news. You hear the weatherman announce "Expect a major heat wave for your region over the next three days. There is a strong possibility of electrical power outages." What would you do next?

1. What **type of disaster** is impacting you and your family?

   ____

2. Is this a **human-made** or **natural disaster**?

   ____

3. How would you **React** to this scenario? Why?

   ____

   ____

4. How would you **Respond** to this scenario? Why?

   ____

   ____

5. Do you **Evacuate** or do you **Shelter in Place**? Why?

   ____

6. What **items will you take with you**? Why?

   ____

   ____

   ____

   ____

7. Name ways to **stay calm** during this scenario?

   ____

   ____

   ____

   ____

Are you prepared?

Practice Scenario 8

It's Sunday Morning, you and your family are getting dressed for church. All of a sudden the house begins to shake you realize an earthquake is taking place. <u>What would you do next?</u>

1. What **type of disaster** is impacting you and your family?

   ____

2. Is this a **human-made** or **natural disaster**?

   ____

3. How would you **React** to this scenario? Why?

   ____

   ____

4. How would you **Respond** to this scenario? Why?

   ____

   ____

5. Do you **Evacuate** or do you **Shelter in Place**? Why?

   ____

6. What **items will you take with you**? Why?

   ____

   ____

   ____

   ____

7. Name ways to **stay calm** during this scenario?

   ____

   ____

   ____

   ____

Notes

Notes

Stay Prepared!

Remember:

Practice Makes Perfect! Keep this book updated and your family prepared!

References

Ascension Parish Emergency Preparedness Guide; www.ascension-caer.org

Bullard, R. D., Mohai, P., Saha, R., & Wright, B. (2008). Toxic wastes and race at twenty: why race still matters after all of these years. Envtl. L., 38, 371.

Bryce, Cyralene P. "Stress Management." *PAHO Library Cataloguing-in-Publication* (2001): 1-138. *PreventionWeb*. Pan American Health Organization. Web. 24 July 2016.

Capolla, Damon. Chapter 2: Preparedness. N.p.: FEMA: Emergency Management Institute, 4 Mar. 2005. DOC.

Cutter, Susan, Bryan J. Boruff, and W. Lynn Shirley. 2003. "Social Vulnerability to Environmental Hazards," Social Science Quarterly 84:2 (June 2003) pp. 242-261.

Chemical Emergencies." Emergency Preparedness and Response. Centers for Disease Control and Prevention, n.d. Web. 24 July 2016. <http://emergency.cdc.gov/chemical/overview.asp>.

Enayati, Amanda. "7 Ways to Manage Stress in a Disaster." CNN. Cable News Network, 31 Oct. 2012. Web. 23 July 2016. <http://www.cnn.com/2012/10/31/health/stress-disaster/>

FEMA. "Talking Points." America's PrepareAthon! Talking Points and Statistics(n.d.): n. pag. 5 Mar. 2014. Web. 23 July 2016.

Lee, Trymaine. "Cancer Alley: Big Industry Big Problem." MSNBC, n.d. Web. <http://www.msnbc.com/interactives/geography-of-poverty/se.html>.

"Natural Disasters." Ready.gov. Department of Homeland Security, n.d. Web. 05 July 2016.

NOAA. "Severe Weather 101." NOAA National Severe Storms Laboratory. The National Severe Storms Laboratory, n.d. Web. 23 July 2016.

"Office of Homeland Security and Emergency Preparedness." Ascension Parish. N.p., n.d. Web. 24 July 2016. <http://ascensionparish.net/>.

Paton, D., & Johnston, D. (2001). Disasters and communities: vulnerability, resilience and preparedness. Disaster Prevention and Management: An International Journal, 10(4), 270-277.

References

"Preparing for a Disaster." International Federation of Red Cross and Red Crescent Societies. IFRC, n.d. Web. 24 July 2016. <http://www.ifrc.org/>.
"Preparedness Attitudes & Behaviors." Earth Institute. Columbia University. National Center for Disaster Preparedness, n.d. Web. 24 July 2016. <http://ncdp.columbia.edu/research/preparedness-attitudes-behaviors/>.

"Preparedness Attitudes & Behaviors." Earth Institute. Columbia University. National Center for Disaster Preparedness, n.d. Web. 24 July 2016. <http://ncdp.columbia.edu/research/preparedness-attitudes-behaviors/>.

PTSD: National Center for PTSD." Traumatic Effects of Specific Types of Disasters -. USDVA, n.d. Web. 23 July 2016. <http://www.ptsd.va.gov/professional/trauma/disaster-terrorism/traumatic-effects-disasters.asp>.

Sutton, Jeannette, and Kathleen Tierney. (2006). *Disaster Preparedness: Concepts, Guidance, and Research.* Boulder: Natural Hazards Center Institute of Behavioral Science, Print.

Tornado Safety and How to Be Safe during a Tornado." Tornado Safety and How to Be Safe during a Tornado. Tornado Chaser, n.d. Web. 24 July 2016. <http://www.tornadochaser.net/safety.html>.

Wingate, Martha S. et al. "Identifying and Protecting Vulnerable Populations in Public Health Emergencies: Addressing Gaps in Education and Training." *Public Health Reports* 122.3 (2007): 422–426. Print.

"Written Testimony of FEMA Administrator Craig Fugate for a House Committee on Transportation and Infrastructure, Subcommittee on Economic Development, Public Buildings and Emergency Management Hearing Titled "Blackout! Are We Prepared to Manage the Aftermath of a Cyber-Attack or Other Failure of the Electrical Grid?"" Homeland Security. N.p., 14 Apr. 2016. Web. 24 July 2016. <https://www.dhs.gov/news/2016/04/14/written-testimony-fema-administrator-house-transportation-and-infrastructure>.

Other Resources

Check out these resources:

https://www.thekapsdisasterhub.com/
https://www.ready.gov/kit
https://www.ready.gov/publications
https://www.usgs.gov
https://hazards.fema.gov
https://www.nssl.noaa.gov/
https://www.redcross.org/
https://www.dhs.gov/prepare-my-family-disaster
https://youth.gov/
https://www.cdc.gov/cpr/readiness/resources.htm
https://echo.epa.gov/report-environmental-violations
https://988lifeline.org/

Footnotes

The materials in this book have been adapted from The Red Cross, The Federal Emergency of Management Agency, and Hazard Mitigation Training for Vulnerable Communities published by Routledge. This book should be used as a guide only. Be sure to connect with your local, regional, and national offices to obtain up-to-date resources when preparing for a disaster. All K.A.P.S. character images were created by Alexis Nunez while the remaining images were downloaded from Canva stock images.

Other Books By Joy Semien

Community Development Books

Hazard Mitigation Training for Vulnerable Communities

Disaster Workbooks

A K.A.P.S. Guide to Preparing for Disasters: The Organization Edition

A K.A.P.S. Guide to Preparing for Disasters: Children's Edition

A K.A.P.S. Guide to Preparing for Disasters: College Edition

A K.A.P.S. Guide to Preparing for Disasters: Caregivers Edition

A K.A.P.S. Guide to Preparing for Disasters: Instructional Guide

A K.A.P.S. Guide to Recover from Disasters

by L.E.E.D. With Joy LLC

Made in the USA
Monee, IL
08 May 2025